THE ONE INCH EQUATION OF THE UNIVERSE

On the same origin of quantum physics and Gravity from Einstein Field Equations and Vacuum energy density scale formalism

BALUNGI FRANCIS

Email: bfrancis@cedat.mak.ac.ug
balungif@gmail.com
Tel: +256(0)703683756

Cover Image Credit: **science photo library**

TABLE OF CONTENTS

DEDICATION

To Wallace Wattles &

My sons Odhran and Leander

LIST OF CONSTANTS

1. *Total mass of ordinary matter in the Universe:*

$$M_U = 1.5 \times 10^{53} kg$$

2. *Electromagnetic coupling constant:*

$$\alpha_e = \frac{1}{137}$$

3. *Gravitational constant:*

$$G = 6.67 \times 10^{-11} Nm^2/kg^2$$

4. *MOND acceleration:*

$$a_o = 1.2 \times 10^{-10} m/s^2$$

5. *Cosmological Constant:*

$$\Lambda = 1.11 \times 10^{-52} m^{-2}$$

6. $\mu = 0.4563 \approx h^2$ *(Where* h *is the value of the Reduced Hubble Constant, It should not be confused with the Planck Constant)*

7. *and* $\beta = 0.1081 \approx \frac{\Omega_b}{h^2}$ *(Where* Ω_b *is the Baryon Density Parameter* $\Omega_b \approx 0.04933$ *and* h *is the value of theReducedHubble Constant)*

Note: $h \equiv \frac{H_o}{(100Km.s^{-1}Mpc^{-1})} = 0.6755$

$\mathrm{H}_o = 2.1928 \times 10^{-18} s^{-1}$ is the Hubble constant (Planck units (2018))

AUTHOR'S NOTE

It has been a long time to reconcile quantum physics and general relativity. To date, no globally accepted theory has been proposed to explain all physical observations. In this work, we reformulated the Einstein Field Equations (EFE) in terms of the Cosmological constant and energy density using the vacuum energy formalism. The proposed equation can be transformed into Maxwell equations in electrodynamics fields with a reduction in the background curvature. From the proposed unified equation we proved that the Cosmological constant follows a square law, that is; the Cosmological constant is proportional to the square of the MOND acceleration and the Planck mass ($\Lambda \propto {a_o}^2 \propto {M_p}^2$). When this result was connected to the calculations of Λ in the LambdaCDM, the following were the findings of the study:

- We found out that the widely popularized MOND cosmological relationships $a_o \sim \sqrt{\frac{\Lambda}{3}}$ and $a_o = \mathrm{H}_o c$ are wrong. Our calculations gives a true relation of

$$a_o = c^2\sqrt{\frac{\Lambda}{62.5131}}$$

, and

$$a_o = \mathrm{H}_o c \sqrt{\frac{3\alpha_e \Omega_\Lambda}{h^2}}$$

- The mass of ordinary matter in the observable universe is calculated without using the critical density, volume or diameter of the Universe. The Results we get agrees with the recent findings by Planck who uses a critical density of $4.08 \times 10^{-28} \frac{kg}{\mathrm{m}^3}$ and volume equals $3.58 \times 10^{80}\mathrm{m}^3$. To differ from this, we still get the same result of

$$M_U = \frac{c^3}{H_o G}\sqrt{\frac{h^2 \alpha_e}{3{\Omega_b}^2 \Omega_\Lambda}} = 1.5 \times 10^{53}\mathrm{kg}$$

- We proved that the fine structure constant is related to the present root-mean-square matter fluctuation $\sigma_8 = 0.8133$, the baryon density parameter Ω_b, the density parameter Ω_Λ, the reduced Hubble parameter h as

$$\alpha_e = 3\left(\frac{{\Omega_b}^2}{h^2}\right)\Omega_\Lambda {\sigma_8}^2 = 0.007299 \sim \frac{1}{137}$$

- The exact equation of the large number 8.4×10^{60} has been found, connecting cosmological parameters and Planck units. Thus, a precise formulation and evidence of Dirac Large Number

Hypothesis has been found, connecting the micro world and the macro world as

$$M_H = \left(m_{pl}{}^2 m_s\right)^{\frac{1}{3}}$$

- We connected the Cosmological constant to the MOND acceleration and the Hubble constant into one Equation

$$\Lambda = \frac{a_o H_o}{c^3} \sqrt{\frac{3\Omega_\Lambda h^2}{\alpha_e}}$$

"Truth is ever to be found in simplicity, and not in the multiplicity and confusion of things"

Is. Newton

1

VACUUM ENERGY DENSITY AND THE COSMOLOGICAL CONSTANT

In this work, we propose that the energy ***density of a vacuum*** of space is proportional to the square of the Force (F) acting on particles in this space. This idea is new and has not been discovered by any physicist somewhere else. Here we propose an alternative approach, which is fully detached from any theoretical model. Instead of expressing the curvature of space time in terms of the energy density of matter present in space time, we branch off to a new proposal and profound idea; that in order to connect General relativity to Quantum mechanics we must have another field equation that explains vacuum energy density in terms of the force that acts on particles in space.

We consider a proposed alternative to quantum gravity, in which the space-time metric is treated as classical, even while matter fields remain quantum. This idea leads to a modification of general relativity at low accelerations.

This model-independent derivation can then be compared with the prediction in ΛCDM, or with any theory of gravity of interest.

Mathematically, the vacuum energy density ρ is expressed as

$$\rho = \frac{\mu F^2}{8\pi\alpha_e \hbar c} = \frac{h^2 F^2}{8\pi\alpha_e \hbar c} \qquad (1)$$

Where ħ is the reduced Planck constant, c is the constant speed of light, α_e is the fine structure constant and $\mu = 0.4563$ is a number determined by experiment, **OR** $\mu = 0.4563 \approx h^2$ *(Where* $h = 0.6755$ *is the value of the Reduced Hubble Constant, It should not be confused with the Planck Constant)*

The energy density describes classical radiation fields, not individual photons

We deal first with a possible concern. It is noted that photons are radiation. It is often stated that radiation is a source of the field, but this statement requires crucial disambiguation in the quantum context, and especially in the light of relativistic level. This is because individual photons (Fock states) are not sources of the gravitational field, even though they reflect "curvature" of the metric due to matter sources. Consider the energy density Eqn1. The standard account notes that the contribution, to the energy density ρ for radiation is with specific constants, ε_0 the permittivity of the vacuum and E denoting the electric field

$$\rho = \frac{h^2 F^2}{8\pi\alpha_e \hbar c}$$

The fine structure constant is given as $\alpha_e = \frac{e^2}{4\pi\varepsilon_o \hbar c}$ which when inserted into the above equation gives a relation between energy density and electric fields as

$$\rho = \frac{h^2 \varepsilon_o}{2}\left(\frac{F}{e}\right)^2$$

From the above given equation we notice that the force (F) per unit elementary charge (e) is the Electric field (E) $\frac{F}{e} = E$,

$$\rho = \frac{h^2 \varepsilon_o}{2} E^2$$

We have therefore derived the energy density of the electromagnetic field.

1.1. The unified equation

According to Einstein Field Equation (EFE) in General Relativity

$$\Lambda = \frac{8\pi G}{c^4}\rho$$

Where Λ denotes the curvature of space time (or the cosmological constant) , G is the Gravitational constant and is the ρ energy density.

According to the energy density (Eqn1), when we substitute Eqn1 for the vacuum energy into the EFE above we rewrite the EFE equation in terms of the Planck Energy $W_p = \sqrt{\frac{\hbar c^5}{G}}$ or Planck length $l_p = \sqrt{\frac{\hbar G}{c^3}}$ as;

$$\Lambda = \frac{G}{\hbar c^5}\left(\frac{h^2}{\alpha_e}\right)F^2 = \frac{1}{W_p{}^2}\left(\frac{h^2}{\alpha_e}\right)F^2$$

OR

$$\Lambda = \frac{1}{(\hbar c)^2}\left(\frac{h^2}{\alpha_e}\right)\left(Fl_p\right)^2 = \frac{1}{(\hbar c)^2}\left(\frac{h^2}{\alpha_e}\right)W_x{}^2 \quad (2)$$

The above simple Equation is the Unified Field Equation that connects General Relativity to Quantum Mechanics and Quantum Electrodynamics into Quantum Gravity.

The above equation in simple terms means that the curvature of space time is related to the square of the work or energy $(W_x = Fl_p)$ required to move a particle through a Planck length when a force F is applied.

1.2. The Cosmological Constant from the Unified equation

This section provides a mathematical proof that a unified equation implies a cosmological constant equation that scales with the square of the MOND acceleration.

In this calculation we consider a Planck-mass-sized black hole remnant or a virtual black hole, which is a hypothetical micro black hole that exists temporarily as a result of a quantum fluctuation of spacetime.

Such Planck-mass black holes may in effect be stable objects if the quantized gaps between their allowed energy levels bar them from emitting Hawking particles or absorbing energy gravitationally like a classical black hole. In such case, they would be weakly interacting massive particles; which in this case could explain dark matter.

Now assume that this virtual black hole accelerates with an associated MOND acceleration ($a_o = 1.2 \times 10^{-10} m/s^2$) when a force (F) is applied to it according to Newton's second law, $F = m_{pl} a_o$ where $m_{pl} = \sqrt{\frac{\hbar c}{G}}$, is the Planck mass then the value of the cosmological constant is

$$\Lambda = \frac{1}{(\hbar c)^2}\left(\frac{h^2}{\alpha_e}\right)\left(m_{pl} a_o l_p\right)^2 = \frac{h^2 {a_o}^2}{\alpha_e c^4}$$

$$\Lambda = \frac{h^2 {a_o}^2}{\alpha_e c^4} \qquad (3)$$

$$\Lambda = 1.11 \times 10^{-52} m^{-2}$$

This value is also in good agreement with the recently observed **Planck (2018)** collaboration cosmological constant values and Planck units for ΩΛ = 0.6889±0.0056 and the

Hubble constant $H_o = 67.66 \pm 0.42\ (km/s)/Mpc = H_o = 2.1928 \times 10^{-18} s^{-1}$

The value of the Cosmological constant is very small simply because the work done to move a Planck mass sized particle accelerating with a MOND acceleration through a distance of Planck length is very small with a value$\left(W_x = m_{pl} a_o l_p = 4.223 \times 10^{-53} J\right)$. Also this value according to the equations derived must be squared to give $\left({W_x}^2 = 1.7834 \times 10^{-105} J^2\right)$

1.4. The relationship between the Hubble Constant and the MOND acceleration

In this section we want to express the MOND acceleration in terms of the Hubble constant in a bid to prove that this acceleration is not only related to Dark matter but it is also an inherent property of space time.

In the standard model of Cosmology known as the ΛCDM model, the cosmological constant has the value of;

$$\Lambda = 3\left(\frac{H_o}{c}\right)^2 \Omega_\Lambda = 1.11 \times 10^{-52} m^{-2}$$

Where $H_o = 2.1928 \times 10^{-18} s^{-1}$ is the Hubble constant and $\Omega_\Lambda = 0.6889$ is the omega Sub lambda or Density Parameter), these are recently observed Planck **(2018)** collaboration cosmological constant values and Planck units for $\Omega\Lambda$ = 0.6889±0.0056 and the Hubble constant $H_o = 67.66 \pm 0.42\ (km/s)/Mpc = \quad H_o = 2.1928 \times 10^{-18} s^{-1}$

The above given value is in agreement with our previously derived value in Eqn3 given as;

$$\Lambda = \frac{h^2 a_o{}^2}{\alpha_e c^4}$$

Equating the above two equations, we can create a relationship between the MOND acceleration and the Hubble constant as,

$$a_o = \mathrm{H}_o c \sqrt{\frac{3\alpha_e \Omega_\Lambda}{h^2}} \qquad (5)$$

$$a_o = 1.2 \times 10^{-10} m/s^2$$

This value is in good agreement with experiment.

1.5. Evidence of Dirac large numbers hypothesis

In this section we deduce that the ratio of the maximum acceleration in Quantum gravity to the MOND acceleration is equal to the ratio of the Planck mass to the minimum mass of a particle and also equal to the ratio of the Planck mass to the mass of the Hubble sphere. The relationship is deduced by equating the Equation for the Cosmological constant Lambda CDM to our unified Eqn3 as

$$\Lambda = 3\left(\frac{\mathrm{H}_o}{c}\right)^2 \Omega_\Lambda = \frac{1}{(\hbar c)^2}\left(\frac{h^2}{\alpha_e}\right)\left(m_{pl}a_o l_p\right)^2$$

Taking square roots on both sides of the equation we have on arranging, the ratio of the ratio of the maximum acceleration in Quantum gravity a_{max} to the MOND acceleration a_o is equal to the ratio of the Planck mass m_{pl} to the minimum mass of a particle in Quantum cosmology (the Hubble scale mass) m_s and also equal to the ratio of the Planck mass m_{pl} to the mass of the Hubble sphere M_H equal to a very large number denoted N.

$$\frac{a_{max}}{a_o} = \frac{m_{pl}}{m_s} = \sqrt{\frac{c^5}{\hbar G \mathrm{H}_o{}^2}} = \frac{m_{pl}}{M_H} = 8.47 \times 10^{60}$$

$$N = 8.47 \times 10^{60}$$

Where;

$a_{max} = \frac{c^2}{l_p}\sqrt{\frac{3\alpha_e \Omega_\Lambda}{h^2}} = 1.0227 \times 10^{51} m/s^2$, is the maximum acceleration in Quantum Gravity.

$m_s = \frac{\mathrm{H}_o \hbar}{\mathrm{c}^2} = 2.57 \times 10^{-69} kg,$ is the minimum mass possible in Quantum Cosmology (the Hubble scale mass). This exceptionally small mass coincides with the minimal measurable gravitational self-energy of a particle which is accepted as minimum quantum of energy $E_{min} = \mathrm{H}_o \hbar = 10^{-33}\mathrm{eV}$. This energy takes substantial place in the estimations of total information and entropy of the observable universe . Thus, the mass m_s seems close to the graviton mass obtained by different methods. The mass m_s is several orders of magnitude smaller than the upper limit of graviton mass, obtained by astrophysical constraints. Therefore, this value doesn't contradict of astrophysical observations.

$M_H = \frac{c^3}{H_0 G} = 1.846 \times 10^{53} kg,$ is the critical mass of the observable universe.

These ratios are very important because they connect cosmological parameters (mass of the Hubble sphere, Hubble scale mass and MOND acceleration and the respective fundamental microscopic properties of matter (Planck mass, Planck length and Planck maximum acceleration in Quantum gravity). In recent quantum gravity models, the Planck units imply quantization of spacetime at extremely short range. Thus, the ratios represent connection between cosmological parameters and quantum properties of spacetime.

Finally the ratio of M_H to m_s gives us a complete and final Equation connecting the micro world to the macro world in this way

$$\frac{M_H}{m_s} = \frac{c^5}{\hbar G {H_o}^2} = \frac{{m_{pl}}^2}{{M_H}^2} = N^2 = 7.174 \times 10^{121}$$

The above calculated large number is also the ratic of Planck density and recent critical density of the universe.

Which when arranged gives an empirical relation

$$M_H{}^3 = m_{pl}{}^2 m_s$$

Connecting the large scale M_H to the small scale m_{pl} and m_s we have

$$M_H = \left(m_{pl}{}^2 m_s\right)^{1/3}$$

Thus, a precise formulation and evidence of Dirac Large number hypothesis has been found connecting the micro world and the macro world.

Since the most of parameters entering equation the ratios are characterized by the maximum or minimum possible values in the universe, it can be roughly claimed that the Dirac LNH shows that the largest and the smallest in the universe are related with a number of the order of $N = 8.47 \times 10^{60}$

1.6. The relationship between the Cosmological Constant and the MOND acceleration

The Hubble constant in the Lambda CDM model above is given as

$$H_o = c\sqrt{\frac{\Lambda}{3\Omega_\Lambda}}$$

Substituting this value into our MOND acceleration (Eqn5) we then get a relation that has baffled scientists for years as

$$a_o = c^2\sqrt{\frac{\alpha_e\Lambda}{h^2}} = c^2\sqrt{\frac{\Lambda}{62.5131}}$$

This value differs widely from the popularized MOND cosmological relation of $a_o \sim \sqrt{\frac{\Lambda}{3}}$. In conclusion According to the relation derived above, this popularized relation is wrong.

2

MASS OF ORDINARY MATTER IN THE UNIVERSE EQUATION

The total mass of ordinary matter in the universe is estimated to be about 1.5 x 10^53 kilograms. This is based on the critical density and the diameter of the observable universe. The values for the critical density, the volume and diameter of the universe can not be measured accurately and they are bound to change with time. This means that they are not constants.

In this section we want to derive the mass of the observable universe without using its volume and diameter. The value we shall get will be in agreement with the Lambda CDM values. To do this, we assume a homogeneous, isotropic, flat Universe and using the fundamental constants for Gravitation (G), Cosmological constant (Λ), and the MOND acceleration (a_o) constant , dimensional analysis gives an estimated mass of

$$M_U = \frac{a_o}{\beta \Lambda G} = \frac{h^2 a_o}{\Omega_b \Lambda G} \qquad (6)$$

Where $\beta = 0.1081$ is a number determined by experiment.

OR $\beta = 0.1081 \approx \frac{\Omega_b}{h^2}$ *(Where* Ω_b *is the Baryon Density Parameter* $\Omega_b \approx 0.04933$ *and* h *is the value of the Reduced Hubble Constant*

The equation for the total mass of ordinary matter in universe contains a term for the MOND acceleration Eqn5. We would like to eliminate this acceleration and thereby create an equation for the mass of matter; we do this by substituting Eqn5 into Eqn6 above

$$M_U = \frac{H_o c}{\Omega_b \Lambda G}\sqrt{3h^2\Omega_\Lambda \alpha_e}$$

But since in the ΛCDM model, the cosmological constant has the value of $\Lambda = 3\left(\frac{H_o}{c}\right)^2 \Omega_\Lambda$, then

$$M_U = \frac{c^3}{H_o G}\sqrt{\frac{h^2\alpha_e}{3{\Omega_b}^2\Omega_\Lambda}} = 1.5 \times 10^{53}\text{kg} \quad (7)$$

This value has been calculated using the cosmological parameters , $h = 0.6755$, $\Omega_b = 0.04933$, $\Omega_\Lambda = 0.6889$ and $H_o = 2.1928 \times 10^{-18} s^{-1}$ (Planck(2018)). The mass of the observable universe hereby derived is independent of the critical density, volume and diameter of the Universe yet it corresponds to the mass that was calculated using the

Planck units in Lambda CDM of 4.8% of the total critical density($0.85 \times 10^{-26}\text{kg/m}^3$) at $\text{H}_\text{o} = 2.176 \times 10^{-18}s^{-1}$ or $4.08 \times 10^{-28}\ kg/\text{m}^3$ and using the commoving distance (radius) of about 46.6 billion light-years, they got a volume equals $3.58 \times 10^{80}\text{m}^3$ and the mass of ordinary matter equals density ($4.08 \times 10^{-28}\ kg/\text{m}^3$) times volume ($3.58 \times 10^{80}\text{m}^3$) or $1.46 \times 10^{-53}\ kg$.

2.1. The Fine structure constant or electromagnetic coupling

We have seen the appearance of the fine structure constant in our equations, the question is, does this constant in any way related to the cosmological parameters?

In this section we derive the fine structure constant by rewriting Eqn7 above in the form

$$M_U = M_H \sqrt{\frac{h^2 \alpha_e}{3{\Omega_b}^2 \Omega_\Lambda}}$$

Where M_H is the mass of the Hubble sphere $M_H = \frac{c^3}{H_o G}$. The Hubble sphere is a spherical region of the universe surrounding an observer beyond which objects recede from that observer at a rate greater than the speed of light.

This implies that the mass of the observed universe is the mass of the Hubble sphere times the square root of some constant of nature related to one of the cosmological parameters we are yet to find out.

The ratio of the observable mass of the universe to the mass of the Hubble sphere is equal to a number as

$$\frac{M_U}{M_H} = \sqrt{\frac{h^2 \alpha_e}{3{\Omega_b}^2 \Omega_\Lambda}} = 0.81387 \qquad (8)$$

The number given above has been purely derived from the known constants given in the theory developed above and nothing new has been added.

By observation, it follows that this number is the same as; $\sigma_8 = 0.8133$ (The present root-mean- square matter fluctuation, averaged over a sphere of radius 8h-1Mpc) and $a^{1/2} = 0.8136$ where a is calculated from; the transition from decelerating to accelerating expansion (the second derivative crossing zero) occurred when

$$a = \left(\frac{\Omega_m}{2\Omega_\Lambda}\right)^{1/3} \sim 0.662$$

The square root of a gives the value $a^{1/2} = 0.8136$.

Also, the S_8 parameter in the ΛCDM model quantifies the amplitude of matter fluctuations in the late universe and is defined as

$$S_8 = \sigma_8 \sqrt{\frac{\Omega_m}{0.3}}$$

Where the constant $\sigma_8 = 0.8133$ (The present root-mean-square matter fluctuation, averaged over a sphere of radius 8h-1Mpc).

Therefore connecting the above density parameters into our mass ratio formula we have

$$\frac{\boldsymbol{M_U}}{\boldsymbol{M_H}} = \sqrt{\frac{\boldsymbol{h^2\alpha_e}}{\boldsymbol{3\Omega_b{}^2\Omega_\Lambda}}} = \left(\frac{\boldsymbol{\Omega_m}}{\boldsymbol{2\Omega_\Lambda}}\right)^{\boldsymbol{1/6}} = \boldsymbol{S_8}\sqrt{\frac{\boldsymbol{0.3}}{\boldsymbol{\Omega_m}}} = \boldsymbol{\sigma_8 = 0.81387}$$

From the above given equality we can get the value of the fine structure constant purely fashioned in the known density parameters as

$$\alpha_e = 3\left(\frac{\Omega_b{}^2}{h^2}\right)\left(\frac{\Omega_m\Omega_\Lambda{}^2}{2}\right)^{1/3} = \frac{0.9\Omega_b{}^2\Omega_\Lambda S_8{}^2}{h^2\Omega_m} = 3\left(\frac{\Omega_b{}^2}{h^2}\right)\Omega_\Lambda\sigma_8{}^2$$

(8)

Where, $\Omega_\Lambda = 0.6889$, $\sigma_8 = 0.8133$, $\Omega_b = 0.04933$, h=0.6755

The electromagnetic coupling constant is therefore given as

$$\boldsymbol{\alpha_e = 3\left(\frac{\Omega_b{}^2}{h^2}\right)\Omega_\Lambda\sigma_8{}^2 = 0.007299\sim 1/137}$$

The above given value is in agreement with experiment. This result is proof that our model is consistent with nature and it also proves that the cosmological parameters herein used are true constants of nature and the Lambda CDM model is valid at the 2018 Planck units.

2.2. The Final Equation; The one Inch Equation of the Universe (A connection between the cosmological constant , the Hubble constant, the MOND acceleration constant, the fine structure constant, the gravitational constant and the density parameters)

Equating Eqn6 to Eqn7, we get a connection between the cosmological constant , the Hubble constant, the MOND acceleration constant, the fine structure constant, the density parameters and the Gravitational constant as,

$$G = \frac{a_o h}{\Lambda M_H}\sqrt{\frac{3\Omega_\Lambda}{\alpha_e}}$$

But Since $M_H = \frac{c^3}{H_o G}$, then the cosmological constant is related to all the constants of nature by

$$\Lambda = \frac{a_o H_o}{c^3}\sqrt{\frac{3\Omega_\Lambda h^2}{\alpha_e}} \qquad (9)$$

ADDITIONAL READINGS

Balungi Francis, (2010) "A hypothetical investigation into the realm of the microscopic and macroscopic universes beyond the standard model" general physics arXiv:1002.2287v1 [physics.gen-ph]

Hawking, Stephen (1975). "Particle Creation by Black Holes". Commun. Math. Phys. 43 (3): 199–220. Bibcode:1975CMaPh..43..199H.

Hawking, S. W. (1974). "Black hole explosions?". Nature.248(5443):30–31.
Bibcode:1974Natur.248...30H.doi:10.1038/248030a0.

Carlo Rovelli (2003) "Quantum Gravity" Draft of the Book Pdf
Some few texts used are from Wikipedia https://creativecommons.org/licenses/by-sa/3.0/
D. N. Page, Phys. Rev. D 13, 198 (1976).

C. Gao and Y.Lu, Pulsations of a black hole in LQG (2012) arXiv:1706.08009v3

A.H. Chamseddine and V.Mukhanov, Non singular black hole (2016) arXiv 1612.05861v1

M.Bojowald and G.M.Paily, A no-singularity scenario in LQG (2012) arXiv: 1206.5765v1

P.Singh, class.Quant.Grav,26,125005(2009), arXiv:0901.2750

P.Singh and F.Vidotto, Phys.Rev, D83,064027(2011) arXiv:1012.1307

C.Rovelli and F.Vidotto, Phy. Rev,111(9) 091303(2013) arXiv:1307.3228v2

M.Bojowald, Initial conditions for a universe, Gravity Research Foundation (2003)

A.Ashtekar, Singularity Resolution in Loop Quantum Cosmology (2008) arXiv:0812.4703v1

J.Brunneumann and T.Thiemann, On singularity avoidance in Loop Quantum Gravity (2005) arXiv:0505032v1

L.Modesto, Disappearence of the Black hole singularity in Quantum gravity (2004) arXiv:0407097v2

Mikhailov, A.A. (1959).Mon. Not. Roy. Astron. Soc.,119, 593.

P. Merat etal.(1974). Astron & Astrophys 32, 471-475

Trempler, R.J. (1956).Helv. Phys. Acta, Suppl.,IV, 106.

Trempler, R.J. (1932). " The deflection of light in the sun's gravitational field "Astronomical society of the pacific 167

Einstein, A. (1916).Ann. d. Phys.,49, 769; (1923).The Principle of Relativity, (translators Perret, W. and Jeffery, G.B.), (Dover Publications, Inc., New York), pp. 109–164.

Von Klüber, H. (1960). InVistas in Astronomy, Vol. 3, pp. 47–77.

K. Hentschel (1992). Erwin Finlay Freundlich and testing Einstein theory of relativity, Communicated by J.D. North

Muhleman, D.O., Ekers, R.D. and Fomalont, E.B. (1970).Phys. Rev. Lett.,24, 1377

Mikhailov, A.A. (1956).Astron. Zh.,33, 912.

Dyson, F.W., Eddington, A.S. and Davidson, C. (1920).Phil. Trans. Roy. sog., A220, 291
Chant, C.A. and Young, R.K. (1924).Publ. Dom. Astron. Obs.,2, 275.

Campbell, W.W. and Trumbler, R.J. (1923).Lick Obs. Bull.,11, 41.

Freundlich, E.F., von Klüber, H. and von Brunn, A. (1931).Abhandl. Preuss. Akad. Wiss. Berlin, Phys. Math. Kl., No.l;Z. Astrophys.,3, 171

Mikhailov, A.A. (1949).Expeditions to Observe the Total Solar Eclipse of 21 September, 1941, (report), (ed. Fesenkov, V.G.), (Publications of the Academy of Sciences, U.S.S.R.), pp. 337–351.

S.P. Martin, in Perspectives on Supersymmetry , edited by G.L. Kane (World Scientific, Singapore, 1998) pp. 1–98; and a longer archive version in hep-ph/9709356; I.J.R. Aitchison, hep-ph/0505105.

Mara Beller, Quantum Dialogue: The Making of a Revolution. University of Chicago Press, Chicago, 2001.

Morrison, Philp: "The Neutrino, scientific American, Vol 194,no.1 (1956),pp.58-68.

R. Haag, J. T. Lopuszanski and M. Sohnius, Nucl. Phys. B88, 257 (1975) S.R. Coleman and J. Mandula, Phys.Rev. 159 (1967) 1251.

H.P. Nilles, Phys. Reports 110, 1 (1984).

P. Nath, R. Arnowitt, and A.H. Chamseddine, Applied N = 1 Supergravity (World Scientific, Singapore, 1984).

S.P. Martin, in Perspectives on Supersymmetry , edited by G.L. Kane (World Scientific, Singapore, 1998) pp. 1–98; and a longer archive version in hep-ph/9709356; I.J.R. Aitchison, hep-ph/0505105.

S. Weinberg, The Quantum Theory of Fields, VolumeIII: Supersymmetry (Cambridge University Press, Cambridge,UK, 2000).

E. Witten, Nucl. Phys. B188, 513 (1981).

S. Dimopoulos and H. Georgi, Nucl. Phys. B193, 150(1981).

N. Sakai, Z. Phys. C11, 153 (1981);R.K. Kaul, Phys. Lett. 109B, 19 (1982).

L. Susskind, Phys. Reports 104, 181 (1984).

L. Girardello and M. Grisaru, Nucl. Phys. B194, 65(1982); L.J. Hall and L. Randall,

Phys. Rev. Lett. 65, 2939(1990);I. Jack and D.R.T. Jones, Phys. Lett. B457, 101 (1999).

For a review, see N. Polonsky, Supersymmetry: Structureand phenomena. Extensions of the standard model, Lect.Notes Phys. M68, 1 (2001).

G. Bertone, D. Hooper and J. Silk, Phys. Reports 405, 279 (2005).

G. Jungman, M. Kamionkowski, and K. Griest, Phys. Reports 267, 195 (1996).

V. Agrawal, S.M. Barr, J.F. Donoghue and D. Seckel,Phys. Rev. D57, 5480 (1998).

N. Arkani-Hamed and S. Dimopoulos, JHEP 0506, 073(2005); G.F. Giudice and A. Romanino, Nucl. Phys. B699, 65(2004) [erratum: B706, 65 (2005)]. July 27, 2006 11:28

en.wikipedia.org/wiki/Supersymmetry - 52k - Cached - Similar pages
en.wikipedia.org/wiki/Grand_unification_theory - 39k - Cached - Similar pages

In cosmology, the Planck epoch (or Planck era), named after Max Planck, is the earliest period of time in the history of the universe, en.wikipedia.org/wiki/**Planck_epoch** - 23k - Cached - Similar pages

L. Shapiro and J. Sol`a, Phys. Lett. B 530, 10 (2002);

E. V.Gorbar and I. L. Shapiro, JHEP 02, 021 (2003); A. M. Pelinson,

L. Shapiro, and F. I. Takakura, Nucl. Phys. B 648, 417 (2003).

A. Starobinsky, Phys. Lett. B 91, 99 (1980).

G. F. R. Ellis, J. Murugan, and C. G. Tsagas, Class. Quant. Grav.21, 233 (2004).

H. V. Peiris et al., Astrophys. J. Suppl. 148, 213 (2003).

D. N. Spergel et al., astro-ph/0603449.

Vilenkin, Phys. Rev. D 32, 2511 (1985).

A. Starobinsky, Pis'ma Astron. Zh 9, 579 (1983).

A.H. Guth, Phys. Rev. D23, 347 (1981).

A.D. Linde, Phys. Lett. B108, 389 (1982); A. Albrecht, P.J.

Steinhardt, Phys.Rev. Lett. 48, 1220 (1982).

A.D. Linde, Phys Lett. B129, 177 (1983).

N. Makino, M. Sasaki, Prog. Theor. Phys. 86, 103 (1991);

D. Kaiser, Phys. Rev.D52, 4295 (1995).

H. Goldberg, Phys. Rev. Lett. 50, 1419 (1983).

E. Kolb and M. Turner, The Early Universe (Addison-Wesley, Reading, MA,1990).

W. Garretson and E. Carlson, Phys. Lett. B 315, 232(1993); H. Goldberg, hep-ph/0003197.

Eddington, A. S., The Internal Constitution of the Stars (Cambridge University Press, England,1926), p. 16

Duncan R .C. & Thompson C., Ap.J.392, L 9 (1992). Thompson , C, Duncan , R .C ., Woods , P., Kouveliotou , C ., Finger , M.H. & van Parad ij s , J .,ApJ, submitted , astro-ph /9908086, (2000).

Schwinger , J .,Phys. Rev.73, 416L (1948)

Carlip, S.: Quantum gravity: a progress report. Rept. Prog. Phys. 64, 885 (2001).arXiv:gr-qc/0108040

Kerr,R.P.: Gravitational field of a spinning mass as an example of algebraically special metrics.

Phys. Rev. Lett. 11, 237–238 (1963)

Bekenstein, J.: Black holes and the second law. Lett. Nuovo Cim. 4, 737–740 (1972)

Bardeen, J.M., Carter, B., Hawking, S.: The four laws of black hole mechanics. Commun.

Math. Phys. 31, 161–170 (1973)

Tolman, R.: Relativity, Thermodynamics, and Cosmology. Dover Books on Physics Series.

Dover Publications, New York (1987)
Oppenheimer, J., Volkoff, G.: On massive neutron cores. Phys. Rev. 55, 374–381 (1939)

Tolman, R.C.: Static solutions of einstein's field equations for spheres of fluid, Phys. Rev. 55,364–373 (1939)

Zel'dovich Y.B.: Zh. Eksp. Teoret. Fiz.41, 1609 (1961)

Bondi, H.: Massive spheres in general relativity. Proc. Roy. Soc. Lond. A281, 303–317 (1964)

Sorkin, R.D., Wald, R.M., Zhang, Z.J.: Entropy of selfgravitating radiation. Gen. Rel. Grav. 1127–1146 (1981)

Newman, E.T., Couch, R., Chinnapared, K., Exton, A., Prakash, A., et al.: Metric of a rotating,charged mass. J. Math. Phys. 6, 918–919 (1965)

Ginzburg, V., Ozernoi, L.: Sov. Phys. JETP 20, 689 (1965)

Doroshkevich, A., Zel'dovich, Y., Novikov I.: Gravitational collapse of nonsymmetric and rotating masses, JETP 49 (1965)

Israel, W.: Event horizons in static vacuum space-times. Phys. Rev. 164, 1776–1779 (1967)

Israel,W.: Event horizons in static electrovac space-times. Commun. Math. Phys. 8, 245–260 (1968)

Loop quantum gravity does provide such a prediction [363, 364], and it disagrees with the semiclassical

Carter, B.: Axisymmetric black hole has only two degrees of freedom. Phys. Rev. Lett. 26, 331–333(1971)

Penrose, R.: Gravitational collapse: the role of general relativity. Riv. Nuovo Cim. 1, 252–276 (1969)

Christodoulou, D.: Reversible and irreversible transformations in black hole physics. Phys. Rev. Lett. 25, 1596–1597 (1970)

Christodoulou, D., Ruffini, R.: Reversible transformations of a charged black hole. Phys. Rev. D4, 3552–3555 (1971)

Hawking, S.: Particle creation by black holes. Commun. Math. Phys. 43, 199–220 (1975)

Klein, O.: Die reflexion von elektronen an einem potential sprung nach der relativistischen dynamik von dirac. Z. Phys. 53, 157 (1929)

Wald, R.M.: General Relativity. The University of Chicago Press, Chicago (1984)

Hawking, S.W.: Black hole explosions. Nature 248, 30–31 (1974)

Hawking, S., Ellis, G.: The large scale structure of space-time. Cambridge University Press, Cambridge (1973)

Carter, B.: Black hole equilibrium states, In Black Holes—Les astres occlus. Gordon and Breach Science Publishers, (1973)

Hawking, S.W.: Gravitational radiation from colliding black holes. Phys. Rev. Lett. 26, 1344–1346 (1971)

Hawking, S.: Black holes in general relativity. Commun. Math. Phys. 25, 152–166 (1972)

Bekenstein, J.: Extraction of energy and charge from a black hole. Phys. Rev. D7, 949–953 (1973)

Bekenstein, J.D.: Black holes and entropy. Phys. Rev. D7, 2333–2346 (1973)

Hawking, S.: Quantum gravity and path integrals. Phys. Rev. D18, 1747–1753 (1978)

Gross, D.J., Perry, M.J., Yaffe, L.G.: Instability of flat space at finite temperature. Phys. Rev. D25, 330–355 (1982)

Unruh, W.G., Wald, R.M.: What happens when an accelerating observer detects a rindler particle. Phys. Rev. D29, 1047–1056 (1984)

Bekenstein, J.D.: Auniversal upper bound on the entropy to energy ratio for bounded systems. Phys. Rev. D23, 287 (1981)

Unruh,W.,Wald, R.M.: Acceleration radiation and generalized second law of thermodynamics. Phys. Rev. D25, 942–958 (1982)

Unruh, W., Wald, R.M.: Entropy bounds, acceleration radiation, and the generalized second law. Phys. Rev. D27, 2271–2276 (1983)

Image : MPI for gravitational physics / W.Benger-zib

Tomilin,K.A., (1999). "Natural Systems Of Units: To The Centenary Aniniversary Of The Planck Systems", 287-296

Sivaram, C. (2007). "What Is Special About the Planck Mass"? arXiv:0707.0058v1

H. Georgi and S.L. Glahow. (1974) "Unity Of All Elementary-Particle Forces". Phys. Rev. Letters 32, 438

Supernova Search Team. Observational evidence from supernovae for an accelerating universe and a cosmological constant. Astron. J.
116, 1009 (1998).

Supernova Cosmology Project. Measurements of Ω and Λ from 42 high redshift supernovae. Astrophys. J. 517, 565 (1999).

Horndeski, G. W. Second-order scalar-tensor field equations in a four-dimensional space. Int. J. Theor. Phys. 10, 363 (1974).

Noller, J. & Nicola, A. Cosmological parameter constraints for Horndeski scalar-tensor gravity. Phys. Rev. D. 99, 103502 (2019).
Koyama, K. Cosmological tests ofmodified gravity. Rept. Prog. Phys. 79, 046902 (2016).

DES collaboration. Dark energy survey year 3 results: constraints on
extensions to ΛCDM with weak lensing and galaxy clustering. Phys. Rev. D. 107, 083504 (2023).

Pogosian, L. et al. Imprints of cosmological tensions in reconstructed gravity. Nat. Astron. 6, 1484 (2022).

Castello, S., Grimm, N. & Bonvin, C. Rescuing constraints on modified gravity using gravitational redshift in large-scale structure. Phys. Rev. D. 106, 083511 (2022).

Bonvin,C. &Pogosian, L.Modified Einstein versus modified Euler for dark matter. Nat. Astron. 7, 1127 (2023).

Song, Y.-S. & Percival, W. J. Reconstructing the history of structure formation using redshift distortions. JCAP 10, 004 (2009).

Blake, C. et al. The WiggleZ dark energy survey: the growth rate of cosmic structure since redshift z=0.9. Mon. Not. R. Astron. Soc. 415, 2876 (2011).

eBOSS collaboration. Completed SDSS-IV extended Baryon oscillation spectroscopic survey:

cosmological implications from two decades of spectroscopic surveys at the apache point observatory. Phys. Rev. D. 103, 083533 (2021).

Tutusaus, I., Sobral-Blanco, D. & Bonvin, C. Combining gravitational lensing and gravitational redshift to measure the anisotropic stress with future galaxy surveys,Phys. Rev. D 107 https://doi.org/10.1103/ physrevd.107.083526 (2023).

DES collaboration. Dark Energy Survey Year 3 results: cosmological constraints from galaxy clustering and weak lensing. Phys. Rev. D. 105, 023520 (2022).

Planck, N. et al. Planck 2018 results. VI. Cosmological parameters. Astron. Astrophys. 641, A6 (2020).

García-García, C. et al. The growth of density perturbations in the last ~10 billion years from tomographic large-scale structure data.

Rozo, D. E. et al. redMaGiC: selecting luminous red galaxies from the DES science verification data. Mon. Not. Roy. Astron. Soc. 461, 1431 (2016).

Marques, G. A. et al. Cosmological constraints from the tomography of DES-Y3 galaxies with CMB lensing from ACT DR4. JCAP 01, 033 (2024).

Chen, S.-F., White, M., DeRose, J. & Kokron, N. Cosmological analysis of three-dimensional BOSS galaxy clustering and Planck CMB lensing cross correlations via Lagrangian perturbation theory. JCAP 07, 041 (2022)

M.J. DUFF, L. B. OKUN, G. VENEZIANO, Trialogue on the number of fundamental constants, J. High Energy Phys., 03,Article ID 023, 2002.

J.D. BARROW, The constants of nature: from alpha to omega, Jonathan Cape, London, 2002.

H. FRITZSCH, The fundamental constants, a mystery of physics, World Scientific, Singapore, 2009.

E. HUBBLE, A relation between distance and radial velocity among extra-galactic nebulae, Proc. Nat. Acad. Sci., 15, pp. 168-173, 1929.
P. J. PEEBLES, Physical cosmology, Princeton Univ. Press, 1971.

D. VALEV, Three fundamental masses derived by dimensional analysis, Am. J. Space Sci., 1, pp. 145-149, 2013.

C. SIVARAM, Cosmological and quantum constraint on particle masses, Am. J. Phys., 50, pp. 279, 1982.

ALFONSO-FAUS, Universality of the self gravitational potential energy of any fundamental particle, Astrophys. Space Sci.,337, pp. 363-365, 2012.

S. LLOYD, Computational capacity of the universe, Phys. Rev. Lett., 88, Article ID 237901, 2002.

A.W. BECKWITH, Energy content of gravitation as a way to quantify both entropy and information generation in the earlyuniverse, J. Modern Phys., 2, pp. 58-61, 2011.

GKIGKITZIS, I. HARANAS, S. KIRK, Number of information and its relation to the cosmological constant resulting fromLandauer's principle, Astrophys. Space Sci., 348, pp. 553-557, 2013.

J. F. WOODWARD, R.J. CROWLEY, W. YOURGRAU, Mach's principle and the rest mass of the graviton, Phys. Rev. D, 11,pp. 1371-1374, 1975.

S. S. GERSHTEIN, A. A. LOGUNOV, M.A. MESTVIRISHVILI, On the upper limit for the graviton mass, Doklady Phys., 43,pp. 293-296, 1998.

D. VALEV, Neutrino and graviton rest mass estimations by a phenomenological approach, arXiv:hep-ph/0507255, 2005.

M.E. ALVES, O.D. MIRANDA, J. C. DE ARAUJO, Can massive gravitons be an alternative to dark energy, arXiv:0907.5190,2009.

A.S. GOLDHABER, M.M. NIETTO, Mass of the graviton, Phys. Rev., D, 9, pp. 1119-1121, 1974.

H. YUKAWA, On the interaction of elementary particles, Proc. Physico-Math. Soc. Japan, 17, pp. 48-57, 1935.

S.W. HAWKING, Black hole explosions?, Nature, 248, pp. 30-31, 1974.

V. VEDRAL, Time, inverse temperature and cosmological inflation as entanglement, arXiv:1408.6965, 2014.

T.M. DAVIS, C.H. LINEWEAVER, Expanding confusion: Common misconceptions of cosmological horizons and the
superluminal expansion of the universe, Publ. Astron. Soc. Australia, 21, pp. 97-109, 2004.

J.R. GOTT III, M. JURIĆ, D. SCHLEGEL, F. HOYLE, M. VOGELEY, M. TEGMARK, N. BAHCALL, J. BRINKMANN, A map of the universe, Astrophys. J., 624, 2, pp. 463-484, 2005.

D. VALEV, Estimations of total mass and density of the observable universe by dimensional analysis, Aerospace Res. Bulgaria,24, pp. 67-76, 2012.

A. BALBI, P. ADE, J. BOCK, J. BORRILL, A. BOSCALERI, P. DE BERNARDIS, P.G. FERREIRA, S. HANANY, V. HRISTOV,A.H. JAFFE, A.T. LEE, Constraints on cosmological parameters from MAXIMA-1, Astrophys. J. Lett., 545, pp. 1-4, 2000.

P. DE BERNARDIS, P.A. ADE, J.J. BOCK, J.R. BOND, J. BORRILL, A. BOSCALERI, K. COBLE, B.P. CRILL, G. DE
GASPERIS, P.C. FARESE, P.G. FERREIRA, A flat Universe from high-resolution maps of the cosmic

microwave background radiation, Nature, 404, pp. 955-959, 2000.

D.N. SPERGEL, L. VERDE, H.V. PEIRIS, E. KOMATSU, M.R. NOLTA, C.L. BENNETT, M. HALPERN, G. HINSHAW,N. JAROSIK, A. KOGUT, M. LIMON, S.S. MEYER, L. PAGE, G.S. TUCKER, J.L. WEILAND, E. WOLLACK,E.L. WRIGHT, First-year Wilkinson microwave anisotropy probe (WMAP) observations: Determination of cosmological parameters, Astrophys. J. Suppl. Ser., 148, pp. 175-194, 2003.

C. L. BENNETT, D. LARSON, J. L. WEILAND, N. JAROSIK, G. HINSHAW, N. ODEGARD, K.M. SMITH, R.S. HILL,B. GOLD, M. HALPERN, E. KOMATSU, M. R. NOLTA, L. PAGE, D.N. SPERGEL, E. WOLLACK, J. DUNKLEY, A. KOGUT, M. LIMON, S. S. MEYER, G. S. TUCKER, E.L. WRIGHT, Nine-year Wilkinson microwave anisotropy probe (WMAP) observations: Final maps and results, Astrophys. J. Suppl. Ser., 208, Article ID 020, 2013.

P.J. MOHR, D.B. NEWELL, B.N. TAYLOR, CODATA recommended values of the fundamental physical constants, J. Phys. Chem. Ref. Data, 45, Article ID 043102 (2016).

V. CANUTO, P. J. ADAMS, S.H. HSIEH, E. TSIANG, Scale-covariant theory of gravitation and astrophysical applications, Phy. Rev. D, 16, pp. 1643-1663, 1977.

Y.K. LAU, The large number hypothesis and Einstein's theory of gravitation, Australian J. Phys., 38, pp. 547-553, 1985.

H.W. PENG, An unification of general theory of relativity with Dirac's large number hypothesis, Commun. Theor. Phys., 42,pp. 703-706, 2004

A. ROGACHEV, The new cosmological model founded on the Scale Covariant Theory of Gravitation and on the Dirac's Large Number Hypothesis. Part 1, arXiv:gr-qc/0606057, 2006.

M. LACHIEZE-REY, E. GUNZIG, The cosmological background radiation, Cambridge Univ. Press, 1999.

E. GREGERSEN, The Universe: A historical survey of beliefs, theories and laws, Rosen Publ. Group, New York, 2010.

Glossary

Absolute space and time—the Newtonian concepts of space and time, in which space is independent of the material bodies within it, and time flows at the same rate throughout the universe without regard to the locations of different observers and their experience of "now."

Acceleration—the rate at which the speed or velocity of a body changes.

Accelerating universe—the discovery in 1998, through data from very distant supernovae, that the expansion of the universe in the wake of the big bang is not slowing down, but is actually speeding up at this point in its history; groups of astronomers in California and Australia independently discovered that the light from the supernovae appears dimmer than would be expected if the universe were slowing down.

Action—the mathematical expression used to describe a physical system by requiring only the knowledge of the initial and final states of the system; the values of the physical variables at all intermediate states are determined by minimizing the action.

Anthropic principle—the idea that our existence in the universe imposes constraints on its properties; an extreme version claims that we owe our existence to this principle.

Asymptotic freedom (or safety)—a property of quantum field theory in which the strength of the coupling between elementary particles vanishes with increasing energy and/or decreasing distance, such that the elementary particles approach free particles with no external forces acting on them; moreover for decreasing energy and/or increasing distance between the particles, the strength of the particle force increases indefinitely.

Baryon—a subatomic particle composed of three quarks, such as the proton and neutron.

Big bang theory—the theory that the universe began with a violent explosion of spacetime, and that matter and energy originated from an infinitely small and dense point.

Big crunch—similar to the big bang, this idea postulates an end to the universe in a singularity.

Binary stars—a common astrophysical system in which two stars rotate around each other; also called a "double star."

Blackbody—a physical system that absorbs all radiation that hits it, and emits characteristic radiation energy depending upon temperature; the concept of blackbodies is useful, among other things, in learning the temperature of stars.

Black hole—created when a dying star collapses to a singular point, concealed by an "event horizon;" the black hole is so dense and has such strong gravity that nothing, including light, can escape it; black holes are predicted by general relativity, and though they cannot be "seen," several have been inferred from astronomical observations of binary stars and massive collapsed stars at the centers of galaxies.

Boson—a particle with integer spin, such as photons, mesons, and gravitons, which carries the forces between fermions.

Brane—shortened from "membrane," a higher-dimensional extension of a onedimensional string.

Cassini spacecraft—NASA mission to Saturn, launched in 1997, that in addition to making detailed studies of Saturn and its moons, determined a bound on the variations of Newton's gravitational constant with time.

Causality—the concept that every event has in its past events that caused it, but no event can play a role in causing events in its past.

Classical theory—a physical theory, such as Newton's gravity theory or Einstein's general relativity, that is concerned with the macroscopic universe, as opposed to theories concerning events at the submicroscopic level such as quantum mechanics and the standard model of particle physics.

Copernican revolution—the paradigm shift begun by Nicolaus Copernicus in the early sixteenth century, when he identified the sun, rather than the Earth, as the center of the known universe.

Cosmic microwave background (CMB)—the first significant evidence for the big bang theory; initially found in 1964 and studied further by NASA teams in 1989 and the early 2000s, the CMB is a smooth signature of microwaves everywhere in the sky, representing the "afterglow"of the big bang: Infrared light produced about 400,000 years after the big bang had redshifted through the stretching of spacetime during fourteen billion years of expansion to the microwave part of the electromagnetic spectrum, revealing a great deal of information about the early universe.

Cosmological constant—a mathematical term that Einstein inserted into his gravity field equations in 1917 to keep the universe static and eternal; although he later regretted this and called it his "biggest blunder," cosmologists today still use the cosmological constant, and some equate it with the mysterious dark energy.

Coupling constant—a term that indicates the strength of an interaction between particles or fields; electric charge and Newton's gravitational constant are coupling constants.

Crystalline spheres—concentric transparent spheres in ancient Greek cosmology that held the moon, sun, planets, and stars in place and made them revolve around the Earth; they were part of the western conception of the universe until the Renaissance.

Curvature—the deviation from a Euclidean spacetime due to the warping of the geometry by massive bodies.

Dark energy—a mysterious form of energy that has been associated with negative pressure vacuum energy and Einstein's cosmological constant; it is hypothesized to explain the data on the accelerating expansion of the universe; according to the standard model, the dark energy, which is spread uniformlythroughout the universe, makes up about 70 percent of the total mass and energy content of the universe.

Dark matter—invisible, not-yet-detected, unknown particles of matter, representing about 30 percent of the total mass of matter according to the standard model; its presence is necessary if Newton's and Einstein's gravity theories are to fit data from galaxies, clusters of galaxies, and cosmology; together, darkmatter and dark energy mean that 96 percent of the matter and energy in the universe is invisible.

Deferent—in the ancient Ptolemaic concept of the universe, a large circle representing the orbit of a planet around the Earth.

Doppler principle or **Doppler effect**—the discovery by the nineteenth-century Austrian scientist Christian Doppler that when sound or light waves are moving toward an observer, the apparent frequency of the waves will be shortened, while if they are moving away from an observer, they will be lengthened; in astronomy this means that the light emitted by galaxies moving away from us is redshifted, and that from nearby galaxies moving toward us is blueshifted.

Dwarf galaxy—a small galaxy (containing several billion stars) orbiting a larger galaxy; the Milky Way has over a dozen dwarf galaxies as companions, including the Large Magellanic Cloud and Small Magellanic Cloud.

Dynamics—the physics of matter in motion.

Electromagnetism—the unified force of electricity and magnetism, discovered to be the same phenomenon by Michael Faraday and James Clerk Maxwell in the nineteenth century.

Electromagnetic radiation—a term for wave motion of electromagnetic fields which propagate with the speed of light—300,000 kilometers per second—and differ only in wavelength; this includes visible light, ultraviolet light, infrared radiation,

X-rays, gamma rays, and radio waves.

Electron—an elementary particle carrying negative charge that orbits the nucleus of an atom.

Eötvös experiments—torsion balance experiments performed by Hungarian Count Roland von Eötvös in the late nineteenth and early twentieth centuries that showed that inertial and gravitational mass were the same to one part in 1011; this was a more accurate determination of the equivalence principle than results achieved by Isaac Newton and, later, Friedrich Wilhelm Bessel.

Epicycle—in the Ptolemaic universe, a pattern of small circles traced out by a planet at the edge of its "deferent" as it orbited the Earth; this was how the Greeks accounted for the apparent retrograde motions of the planets.

Equivalence principle—the phenomenon first noted by Galileo that bodies falling in a gravitational field fall at the same rate, independent of their weight and composition; Einstein extended the principle to show that gravitation is identical (equivalent) to acceleration.

Escape velocity—the speed at which a body must travel in order to escape a strong gravitational field; rockets fired into orbits around the Earth have calculated escape velocities, as do galaxies at the periphery of galaxy clusters.

Ether (or aether)—a substance whose origins were in the Greek concept of "quintessence," the ether was the medium through which energy and matter moved, something more than a vacuum and less than air; in the late nineteenth century the Michelson-Morley experiment disproved the existence of the ether.

Euclidean geometry—plane geometry developed by the third-century bc Greek mathematician Euclid; in this geometry, parallel lines never meet.

Fermion—a particle with half-integer spin, like protons and electrons, that make up matter.

Field—a physical term describing the forces between massive bodies in gravity and electric charges in electromagnetism; Michael Faraday discovered the concept of field when studying magnetic conductors.

Field equations—differential equations describing the physical properties of interacting massive particles in gravity and electric charges in electromagnetism; Maxwell's equations for electromagnetism and Einstein's equations of gravity are prominent examples in physics.

Fifth force or **"skew" force**—a new force in MOG that has the effect of modifying gravity over limited length scales; it is carried by a particle with mass called the phion.

Fine-tuning—the unnatural cancellation of two or more large numbers involving an absurd number of decimal places, when one is attempting to explain a physical phenomenon; this signals that a true understanding of the physical phenomenon has not been achieved.

Fixed stars—an ancient Greek concept in which all the stars were static in the sky, and moved around the Earth on a distant crystalline sphere.

Frame of reference—the three spatial coordinates and one time coordinate that an observer uses to denote the position of a particle in space and time.

Galaxy—organized group of hundreds of billions of stars, such as our Milky Way.

Galaxy cluster—many galaxies held together by mutual gravity but not in as organized a fashion as stars within a single galaxy.

Galaxy rotation curve—a plot of the Doppler shift data recording the observed velocities of stars in galaxies; those stars at the periphery of giant spiral galaxies are observed to be going faster than they "should be" according to Newton's and Einstein's gravity theories.

General relativity—Einstein's revolutionary gravity theory, created in 1916 from a mathematical generalization of his theory of special relativity; it changed our concept of gravity from Newton's universal force to the warping of the geometry of spacetime in the presence of matter and energy.

Geodesic—the shortest path between two neighboring points, which is a straight line in Euclidian geometry, and a unique curved path in four-dimensional spacetime.

Globular cluster—a relatively small, dense system of up to millions of stars occurring commonly in galaxies.

Gravitational lensing—the bending of light by the curvature of spacetime; galaxies and clusters of galaxies act as lenses, distorting the images of distant bright galaxies or quasars as the light passes through or near them.

Gravitational mass—the active mass of a body that produces a gravitational force on other bodies.

Gravitational waves—ripples in the curvature of spacetime predicted by general relativity; although any accelerating body can produce gravitational radiation or waves, those that could be detected by experiments would be caused by cataclysmic cosmic events.

Graviton—the hypothetical smallest packet of gravitational energy, comparable to the photon for electromagnetic energy; the graviton has not yet been seen experimentally.

Group (in mathematics)—in abstract algebra, a set that obeys a binary operation that satisfies certain axioms; for example, the property of addition of integers makes a group; the branch of mathematics that studies groups is called group theory.

Hadron—a generic word for fermion particles that undergo strong nuclear interactions.

Hamiltonian—an alternative way of deriving the differential equations of motion for a physical system using the calculus of variations; Hamilton's principle is also called the "principle of stationary action" and was originally formulated by Sir William Rowan Hamilton for classical mechanics; the principle applies to classical fields such as the gravitational and electromagnetic fields, and has had important applications in quantum mechanics and quantum field theory.

Homogeneous—in cosmology, when the universe appears the same to all observers, no matter where they are in the universe.

Inertia—the tendency of a body to remain in uniform motion once it is moving, and to stay at rest if it is at rest; Galileo discovered the law of inertia in the early seventeenth century.

Inertial mass—the mass of a body that resists an external force; since Newton, it has been known experimentally that inertial and gravitational mass are equal; Einstein used this equivalence of inertial and gravitational mass to postulate his equivalence principle, which was a cornerstone of his gravity theory.

Inflation theory—a theory proposed by Alan Guth and others to resolve the flatness, horizon, and homogeneity problems in the standard big bang model;

the very early universe is pictured as expanding exponentially fast in a fraction of a second.

Interferometry—the use of two or more telescopes, which in combination create a receiver in effect as large as the distance between them; radio astronomy in particular makes use of interferometry.

Inverse square law—discovered by Newton, based on earlier work by Kepler, this law states that the force of gravity between two massive bodies or point particles decreases as the inverse square of the distance between them.

Isotropic—in cosmology, when the universe looks the same to an observer, no matter in which direction she looks.

Kelvin temperature scale—designed by Lord Kelvin (William Thomson) in the mid-1800s to measure very cold temperatures, its starting point is absolute zero, the coldest possible temperature in the universe, corresponding to –273.15 degrees Celsius; water's freezing point is 273.15K (0°C), while its boiling point is 373.15K (100°C).

Lagrange points—discovered by the Italian-French mathematician Joseph-Louis Lagrange, these five special points are in the vicinity of two orbiting masses where a third, smaller mass can orbit at a fixed distance from the larger masses; at the Lagrange points, the gravitational pull of the two large masses precisely equals the centripetal force required to keep the third body, such as a satellite, in a bound orbit; three of the Lagrange points are unstable, two are stable.

Lagrangian—named after Joseph-Louis Lagrange, and denoted by L, this mathematical expression summarizes the dynamical properties of a physical system; it is defined in classical mechanics as the kinetic energy T minus the potential energy V; the equations of motion of a system of particles may be derived from the Euler-Lagrange equations, a family of partial differential equations.

Light cone—a mathematical means of expressing past, present, and future space and time in terms of spacetime geometry; in four-dimensional Minkowski spacetime, the light rays emanating from or arriving at an event separate spacetime into a past cone and a future cone which meet at a point corresponding to the event.

Lorentz transformations—

mathematical transformations from one inertial frame of reference to another such that the laws of physics remain the same; named after Hendrik Lorentz, who developed them in 1904, these transformations form the basic mathematical equations underlying special relativity.

Mercury anomaly—a phenomenon in which the perihelion of Mercury's orbit advances more rapidly than predicted by Newton's equations of gravity; when Einstein showed that his gravity theory predicted the anomalous precession, it was the first empirical evidence that general relativity might be correct.

Meson—a short-lived boson composed of a quark and an antiquark, believed to bind protons and neutrons together in the atomic nucleus.

Metric tensor—mathematical symmetric tensor coefficients that determine the infinitesimal distance between two points in spacetime; in effect the metric tensor distinguishes between Euclidean and non-Euclidean geometry.

Michelson-Morley experiment—1887 experiment by Albert Michelson and Edward Morley that proved that the ether did not exist; beams of light traveling in the same direction, and in the perpendicular direction, as the supposed ether showed no difference in speed or arrival time at their destination.

Milky Way—the spiral galaxy that contains our solar system.

Minkowski spacetime—the geometrically flat spacetime, with no gravitational effects, first described by the Swiss mathematician Hermann Minkowski; it became the setting of Einstein's theory of gravity.

MOG—my relativistic modified theory of gravitation, which generalizes Einstein's general relativity; MOG stands for "Modified Gravity."

MOND—a modification of Newtonian gravity published by Mordehai Milgrom in 1983; this is a nonrelativistic phenomenological model used to describe rotational velocity curves of galaxies; MOND stands for "Modified Newtonian Dynamics."

Neutrino—an elementary particle with zero electric charge; very difficult to detect, it is created in radioactive decays and is able to pass through matter almost undisturbed; it is considered to have a tiny mass that has not yet been accurately measured.

Neutron—an elementary and electrically neutral particle found in the atomic nucleus, and having about the same mass as the proton.

Nuclear force—another name for the strong force that binds protons and neutrons together in the atomic nucleus.

Nucleon—a generic name for a proton or neutron within the atomic nucleus.

Neutron star—the collapsed core of a star that remains after a supernova explosion; it is extremely dense, relatively small, and composed of neutrons.

Newton's gravitational constant—the constant of proportionality, G, which occurs in the Newtonian law of gravitation, and says that the attractive force between two bodies is proportional to the product of their masses and inversely proportional to the square of the distance between them; its numerical value is: G = 6.67428 ± 0.00067 x 10–11 m3 kg–1 s–2.

Nonsymmetric field theory (Einstein)—a mathematical description of the geometry of spacetime based on a metric tensor that has both a symmetric part and an antisymmetric part; Einstein used this geometry to formulate a unified field

theory of gravitation and electromagnetism.

Nonsymmetric Gravitation Theory (NGT)—my generalization of Einstein's purely gravitation theory (general relativity) that introduces the antisymmetric field as an extra component of the gravitational field; mathematically speaking, the nonsymmetric field structure is described by a non-Riemannian geometry.

Parallax—the apparent movement of a nearer object relative to a distant background when one views the object from two different positions; used with triangulation for measuring distances in astronomy.

Paradigm shift—a revolutionary change in belief, popularized by the philosopher Thomas Kuhn, in which the majority of scientists in a given field discard a traditional theory of nature in favor of a new one that passes the tests of experiment and observation; Darwin's theory of natural selection, Newton's gravity theory, and Einstein's general relativity all represented paradigm shifts.

Parsec—a unit of astronomical length equal to 3.262 light years.

Particle-wave duality—the fact that light in all parts of the electromagnetic spectrum (including radio waves, X-rays, etc., as well as visible light) sometimes acts like waves and sometimes acts like particles or photons; gravitation may be similar, manifesting as waves in spacetime or graviton particles.

Perihelion—the position in a planet's elliptical orbit when it is closest to the sun.

Perihelion advance—the movement, or changes, in the position of a planet's perihelion in successive revolutions of its orbit over time; the most dramatic perihelion advance is Mercury's, whose orbit traces a rosette pattern.

Perturbation theory—a mathematical method for finding an approximate solution to an equation that cannot be solved exactly, by expanding the solution in a series in which each successive term is smaller than the preceding one.

Phion—name given to the massive vector field in MOG; it is represented both by a boson particle, which carries the fifth force, and a field.

Photoelectric effect—the ejection of electrons from a metal by X-rays, which proved the existence of photons; Einstein's explanation of this effect in 1905 won him the Nobel Prize in 1921; separate experiments proving and demonstrating the existence of photons were performed in 1922 by Thomas Millikan and Arthur Compton, who received the Nobel Prize for this work in 1923 and 1927, respectively.

Photon—the quantum particle that carries the energy of electromagnetic waves; the spin of the photon is 1 times Planck's constant h.

Pioneer 10 and 11 spacecraft—launched by NASA in the early 1970s to explore the outer solar system, these spacecraft show a small, anomalous acceleration as they leave the inner solar system.

Planck's constant (h)—a fundamental constant that plays a crucial role in quantum mechanics, determining the size of quantum packages of energy such as the photon; it is named after Max Planck, a founder of quantum mechanics

Principle of general covariance—Einstein's principle that the laws of physics remain the same whatever the frame of reference an observer uses to measure physical quantities.

Principle of least action—more accurately the principle of *stationary* action, this variational principle, when applied to a mechanical system or a field theory, can be used to derive the equations of motion of the system; the credit for discovering the principle is given to Pierre-Louis Moreau Maupertius but it may have been discovered independently by Leonhard Euler or Gottfried Leibniz.

Proton—an elementary particle that carries positive electrical charge and is the nucleus of a hydrogen atom.

Ptolemaic model of the universe—the predominant theory of the universe until the Renaissance, in which the Earth was the heavy center of the universe and all other heavenly bodies, including the moon, sun, planets, and stars, orbited around it; named for Claudius Ptolemy.

Quantize—to apply the principles of quantum mechanics to the behavior of matter and energy (such as the electromagnetic or gravitational field energy); breaking down a field into its smallest units or packets of energy.

Quantum field theory—the modern relativistic version of quantum mechanics used to describe the physics of elementary particles; it can also be used in nonrelativistic fieldlike systems in condensed matter physics.

Quantum gravity—attempts to unify gravity with quantum mechanics.

Quantum mechanics—the theory of the interaction between quanta (radiation) and matter; the effects of quantum mechanics become observable at the submicroscopic distance scales of atomic and particle physics, but macroscopic quantum effects can be seen in the phenomenon of quantum entanglement.

Quantum spin—the intrinsic quantum angular momentum of an elementary particle; this is in contrast to the classical orbital angular momentum of a body rotating about a point in space.

Quark—the fundamental constituent of all particles that interact through the strong nuclear force; quarks are fractionally charged, and come in several varieties; because they are confined within particles such as protons and neutrons, they cannot be detected as free particles.

Quasars—"quasi-stellar objects," the farthest distant objects that can be detected with radio and optical telescopes; they are exceedingly bright, and are believed to be young, newly forming galaxies; it was the discovery of quasars in 1960 that quashed the steady-state theory of the universe in favor of the big bang.

Quintessence—a fifth element in the ancient Greek worldview, along with earth, water, fire and air, whose purpose was to move the crystalline spheres that supported the heavenly bodies orbiting the Earth; eventually this concept became known as the "ether," which provided the *something* that bodies needed to be in contact with in order to move; although Einstein's special theory of relativity dispensed with the ether, recent explanations of the acceleration of the universe call the varying negative pressure vacuum energy "quintessence."

Redshift—a useful phenomenon based on the Doppler principle that can indicate whether and how fast bodies in the universe are receding from an observer's position on Earth; as galaxies move rapidly away from us, the frequency of the wavelength of their light is shifted toward the red end of the electromagnetic spectrum; the amount of this shifting indicates the distance of the galaxy.

Riemann curvature tensor—a mathematical term that specifies the curvature of four-dimensional spacetime.

Riemannian geometry—a non-Euclidean geometry developed in the mid-nineteenth century by the German mathematician George Bernhard Riemann that describes curved surfaces on which parallel lines *can* converge, diverge, and even intersect, unlike Euclidean geometry; Einstein made Riemannian geometry the mathematical formalism of general relativity.

Satellite galaxy—a galaxy that orbits a host galaxy or even a cluster of galaxies.

Scalar field—a physical term that associates a value without direction to every point in space, such as temperature, density, and pressure; this is in contrast to a vector field, which has a direction in space; in Newtonian physics or in electrostatics, the potential energy is a scalar field and its gradient is the vector force field; in quantum field theory, a scalar field describes a boson particle with spin zero.

Scale invariance—distribution of objects or patterns such that the same shapes and distributions remain if one increases or decreases the size of the length scales or space in which the objects are observed; a common example of scale invariance

is fractal patterns.

Schwarzschild solution—an exact spherically symmetric static solution of Einstein's field equations in general relativity, worked out by the astronomer Karl Schwarzschild in 1916, which predicted the existence of black holes.

Self-gravitating system—a group of objects or astrophysical bodies held together by mutual gravitation, such as a cluster of galaxies; this is in contrast to a "bound system" like our solar system, in which bodies are mainly attracted to and revolve around a central mass.

Singularity—a place where the solutions of differential equations break down; a spacetime singularity is a position in space where quantities used to determine the gravitational field become infinite;

such quantities include the curvature of spacetime and the density of matter.

Spacetime—in relativity theory, a combination of the three dimensions of space with time into a four-dimensional geometry; first introduced into relativity by Hermann Minkowski in 1908.

Special theory of relativity—Einstein's initial theory of relativity, published in 1905, in which he explored the "special" case of transforming the laws of physics from one uniformly moving frame of reference to another; the equations

of special relativity revealed that the speed of light is a constant, that objects appear contracted in the direction of motion when moving at close to the speed of light, and that E = mc2, or energy is equal to mass times the speed of light squared.

Spin—see quantum spin.

String theory—a theory based on the idea that the smallest units of matter are not point particles but vibrating strings; a popular research pursuit in physics for two decades, string theory has some attractive mathematical features, but has yet to make a testable prediction.

Strong force—see nuclear force.

Supernova—spectacular, brilliant death of a star by explosion and the release of heavy elements into space; supernovae type 1a are assumed to have the same intrinsic brightness and are therefore used as standard candles in estimating cosmic distances.

Supersymmetry—a theory developed in the 1970s which, proponents claim, describes the most fundamental spacetime symmetry of particle physics: For every boson particle there is a supersymmetric fermion partner, and for every fermion there exists a supersymmetric boson partner; to date, no supersymmetric particle partner has been detected.

Tully-Fisher law—a relation stating that the asymptotically flat rotational velocity of a star in a galaxy, raised to the fourth power, is proportional to the mass or luminosity of the galaxy.

Unified theory (or unified field theory)—a theory that unites the forces of nature; in Einstein's day those forces consisted of electromagnetism and gravity; today the weak and strong nuclear forces must also be taken into account, and perhaps someday MOG's fifth force or skew force will be included; no one has yet discovered a successful unified theory.

Vacuum—in quantum mechanics, the lowest energy state, which corresponds to the vacuum state of particle physics; the vacuum in modern quantum field theory is the state of perfect balance of creation and annihilation of particles and antiparticles.

Variable Speed of Light (VSL) cosmology—an alternative to inflation theory, in which the speed of light was much faster at the beginning of the universe than it is today; like inflation, this theory solves the horizon and flatness problems in the very early universe in the standard big bang model.

Vector field—a physical value that assigns a field with the position and direction of a vector in space; it describes the force field of gravity or the electric and magnetic force fields in James Clerk Maxwell's field equations.

Virial theorem—a means of estimating the average speed of galaxies within galaxy clusters from their estimated average kinetic and potential energies.

Vulcan—a hypothetical planet predicted by the nineteenth-century astronomer Urbain Jean Joseph Le Verrier to be the closest orbiting planet to the sun; the presence of Vulcan would explain the anomalous precession of the perihelion of Mercury's orbit; Einstein later explained the anomalous precession in general relativity by gravity alone.

Weak force—one of the four fundamental forces of nature, associated with radioactivity such as beta decay in subatomic physics; it is much weaker than the strong nuclear force but still much stronger than gravity.

X-ray clusters—galaxy clusters with large amounts of extremely hot gas within them that emit X-rays; in such clusters, this hot gas represents at least twice the mass of the luminous stars.

www.ingramcontent.com/pod-product-compliance
Lightning Source LLC
LaVergne TN
LVHW090430210926
848480LV00002B/761
9798231196579